I0156666

SILT

SILT

PROSE POEMS

BY AURORA LEVINS MORALES

© 2019 Aurora Levins Morales

All rights reserved. No part of this book may be reproduced or utilized in any form or by any means, electronic or mechanical, including photocopying, recording or any information storage and retrieval system without permission in writing from the author or her authorized representative.

Reprint requests:
The Permissions Company at *permdude@gmail.com*

Cover art by Aurora Levins Morales.
Book design by Wordzworth Ltd.

ISBN 978-0-9836831-2-4

Library of Congress Control Number:2019905380

Palabrera Press, Petaluma, CA

¿Quien dijo que todo está perdido?
Yo vengo a ofrecer mi corazón.
Tanta sangre que se llevó el río.
Yo vengo a ofrecer mi corazón.

Who says that everything is lost?
I come to offer you my heart.
So much blood that the river swept away.
I come to offer you my heart.

FITO PAEZ

For my parents, Dick Levins and Rosario Morales who raised me at
the confluence of art, ecology and justice in a house
lined with books and filled with lizards.

And for the generations to come.
May our lives fertilize theirs.

Contents

TWO
Mar

THREE
Wind

FOUR
Clouds

Preface: Silt of Each Other

It doesn't matter what his name was. He could have been any one of the white, propertied men who were the muscle of that fast-beating heart of US expansionism in the late19th century, full of the exhilarating conviction that they were manifestly destined to rule, were the pinnacle of creation, and that anything that did not belong to them was wasted. To dominate nature was to do God's unfinished work, to straighten out His crooked rivers and put them to lighting bulbs and watering crops, to plow up His useless prairies and irrigate His unproductive deserts so they could be planted with mathematically laid out, maximum yield rows of highly marketable crops, to the enrichment of His chosen creatures, the banker, the agribusinessman and the railroad baron. God and natural law stood united behind him and his kind, in support of American Empire.

The internal frontier had been abolished and every nook and cranny of arable Indian land had been stolen. (Who was to know those barren reservation wastes would turn out to have coal and oil and uranium under their rough dry exteriors?) The glittering oceans beckoned east and west with their delectable morsels of tropical conquest. Today it was Cuba on the menu, and the men of means were salivating. It was one of the last gems in the Spanish crown, and they wanted it, they needed it, they must have as soon as possible. So, they drafted reasons for seizing it.

The man in the waistcoat said that the reason was mud. He said that the great river carried silt from American fields, American rocks, American leaves and twigs, far out of its muddy mouth, deep into the Caribbean sea, and that it was these chips of Minnesota granite, fallen pine debris of Wisconsin and Illinois, white kaolin and black shale of Montana, crumbled river bluff from Iowa and Missouri, red clay and sand of Mississippi, Arkansas and Louisiana, sludge of long dead catfish and northern pike, that had piled up over the eons and made Cuba, literally, into US soil.

The islands where I was born rose in fire, out of a place of tilt and slip, strike and jolt, a crack of transformation in the great plates of the earth's skin; a place where one slab of rock shoves its way under another, heaving immense angles of thick wet limestone into undersea mountain ranges whose crowns rise up in foam and salt, to be decked with palms and egrets.

They rose in serpentine and basalt, and the slowly built sedimentary stripes of compressed ocean bed; in the bones of billions of tiny coral creatures, pressed by the weight of water into banks of dolomite that the surf wore smooth, that the sun heated and cracked, that rainwater carved full of caves, seeping through filtering porous beds to make deep sweet aquifers and habitat for seed and spore. They rose in the slow conversion of plant matter to clay, and clay back into plants, into airy bromeliads, towering trunks with splendid crowns, tiny mosses, extravagant bloom, rattling seed pods, fruits and tubers searingly poisonous, exquisitely flavored, earthily nourishing.

These islands did not rise from the trash heaps of the continental river, were not made out of leftover Midwest. They are their own creation, dripping with saltwater and honeysuckle.

And sediment has no nationality. Sediment drifts from place to place, on currents of water and air, on muskrat fur and the feathers of Indigo Buntings. It travels without passports, visas, or allegiances. If there is Minnesota dust ground into fine powder by its long journey through wild currents and sandbars; fragments of Canadian glacial rock whirled along the bed of the Missouri into the common flood; grains of ancient lake bottom that have been swept out to sea and fallen at last into drifts of silken mud, pressed into shale when our ancestors were still tree rats, and in an age long before borders, lifted and tipped into the side of some Caribbean valley, why, each one of us walks through the day with miniscule bits of old African DNA twirling in the mitochondria that fuel our cells. Caribbean pollen of tropical forests brushed from the feathers of migratory birds has also sifted down through the fragrance of fermenting pine needles, into the lake bed of the River's source, mingling, under the rose green flickering of the northern lights, with the ashes of volcanic eruptions half a world away.

What are we all but the silt of each other? Every molecule of oxygen we breathe has been breathed a billion times before by every other set of human lungs, has crossed the panting tongues of dogs and leopards and small jeweled snakes, fueled the tiny hearts of hummingbirds and the large hearts of gorillas, been transformed in the green veins of plants, passed in and out of clouds and rain and waterfalls, risen up from oceans into the planetary sky. To whom does it belong?

We are the mineral residues of distant stars, carbons formed within their blazing cores, continually rearranging ourselves in an endless genetic shuffle of dazzling forms. Microscopic flakes of our skin lie scattered on Himalayan snow and in the bed of the Amazon.

We cannot be owned. We cannot own each other. And there is no such thing as a particle of US soil.

But there *is* the root web of eco-history, a dense mat of relations thick as mangrove, between the great inland river and the islands that girdle the small bright sea.

Birds and pebbles, seeds and sorrows, languages and tools, trees and music, bloodshed and rebellion, sugar and iron, people and their stories have traveled the waters upriver and down, windward and leeward, and left their mark.

These poems are drawn from the real story of river and island silt, the residues of landscapes and peoples, species and cultures, that shift and change and charge each other with minerals and meaning, moving along a pathway made out of water and mud and our own feet walking. I dip my fingers in its rich colors and paint my face. I dip my fingers and write my own manifest destiny: we are meant to be fearless, we are meant to recognize each other and rejoice, we are meant to be free.

Introduction: New Orleans

I came to a city at the crossroad of earth and water, where the Great River fans out in a thousand fingers to meet the small bright Sea. I came carrying my own dead. Mis muertx.

There is an open pit mine in this place, where stories are shoveled out of the half-flooded soil and taken away, to shine in private collections housed on solid ground. I didn't bring a shovel. I came with my own stories, my handful of seashells, my granito de arroz, my island eyes, to find the confluence of our lives. I came to Bulbancha, where many words meet, to add my story to the shell mound.

I came carrying my drowned country in my hands. I came in the wake of storms that were not brewed at sea. I came from a land on fire. I did not come to sit shiva. I did not come to sit at the wake. I came to add my song to the weave.

I came carrying my mother Sari who taught me the brujeria of her life, buzzing with questions: taught me how to hold a pencil and lay words onto paper, how to track the course of a hurricane across a map by lamplight, how to brew tea from night blossoming jasmine, to know the names of all the colors (cerulean, viridian, rose madder, burnt umber) and make them my compass, to turn toward difficulty and chart my course, to sew a swift, well anchored seam, to smell the flaw in the argument, the problem with the plan, the name of the dust brushed under the rug, but also to smell the first hint of dawn, and the promise of rain.

I came carrying my father DIck who taught me the ciencia of his life, drawing equations in wet sand and illustrated birthday books in rhyme: who taught me endless webs of connection, and all the words of Solidarity Forever, who showed me the wings of fruit flies, their ruby eyes, the banded shells of snails, the poetry of imprisoned radicals—impossible blooming, who laid his hand on my head and said think, trust your mind, who told me there are no bad people, only bad decisions and chanted complexity, complexity, the truth is the whole.

I carry my mother in my bones, my father in the chambers of my heart, though she died of cancer riddled bone marrow and he died of a muscle that could barely move his blood.

I came to let this place rise through the soles of my feet, through my bones and my heart and turn into the water songs of a traveling Caribbean Jew.

ONE

Manantiales

Atabey :: Ochun

Manantiales : Bimini (First water)

On the mountain where I was born, there are seven springs. They rise up between chunks of serpentine and banks of red clay in a place called Indiera, meaning place of the Indians, my Taino ancestors, in the western mountains of an island my people called Borikén, before it was named after a holy person revered by the invaders who came to pillage, and then named "rich port" after the pillage they took.

The springs flow out of the mountainside, out of the pooling of rain-water that pours from the sky each autumn, and trickles downhill, following the curves and crevices of the land until it joins rivers with names, and finds its way into the deep, wild channel to the west, where strong, ever-changing currents rush through deep undersea canyons, and the sea bed trembles.

But high in the rain hung mountains, where clouds move through valleys and wreath the mountaintops, my ancestors linger in the mist, in the forests, in the clay.

I wake before dawn to the cries of owls calling all around me in the trees, and the voices of the ancestors slipping through the space between sleep and waking, summoning me to listen. They say the mountain and its forests and waters are a reservoir of souls, a memory archive of the old ones, a place to protect and defend. They say I must protect water. They say each morning when I rise, no matter where I am, I must drink from these springs and let their water flow through me and out of my mouth as words.

Rain

There is no beginning to this story. We begin midstream. We begin in a gush of water from our mother's bodies. In our first moist breath. The makers of landmarks and destinations, the holders of deeds, argue about where the River is born. In what exact spot does it really begin, which trickle into which stream, into which tributary, how many gallons, how fast, how wide, how deep? They say it begins in a clear, cold northern lake, Omashkoozo-zaaga'igan, a shallow glacial hollow in the bedrock, and travels this course or that along the ancient claw marks of ice. But it is rain that gathers there and begins to flow. It is the snow that fell in drifts from heavy clouds and melted in the sun. The headwaters of the Mississippi River are in the sky.

The sea has no headwater. It is all belly. The sun beats down until the surface of the salt water turns into dimpled gold. It steams up into the sky, becoming low, white clouds of the Caribbean, held close because the upper layers of air keep them so, or is swept into the great spiral storms that come spinning in herds across the ocean from Africa, or is blown northward in streamers and rags of vapor, on the arc of the winds. Steam meets the chill upper air, meets the cool shadowy breath of trees, tumbles and darkens and condenses, and the water in the sky returns in spatters, sheets, pillars of rain, pouring sweet into salt, pockmarking endless miles of rolling waves, and blows ashore, drenching the islands, drenching the continent, returning, always returning.

Water flows in and out of red veins and green, making a tracery of drops spangling the globe, this wet planet, this single organism we are. We are made of water's dance. Everything else is residue, molecules of matter and spirit moving from place to place. Everything else is silt.

2

Windblown water beats on the metal roof of the house where I am still a child, drumming so loud we have to shout. It comes on the vientos alisios, the sea winds, the trade winds, always blowing from the northeast. The sea winds shepherd the great storms with their trailing whips of dark cloud, their spinning hearts, their debris of broken branches, torn paper, pieces of houses. They bring red dust of Mali, blow tiny particulate matter into our lungs, sticky traces of pesticides and truck exhaust into the crevices of living reefs. They bring invaders blown off course from Iberia, lost in a dream of India and Cathay, hacking our worlds apart like the husk around a kernel of gold. Not corn, but money.

The winds that run low, closest to the moist earth and the crested sea, blow toward the equator. The winds that run high in the upper air race away toward the poles. The winds of the islands come slant across the map and bring water. Billions of tiny tree frogs call out as the humidity rises. The old ones say they are calling for their lost mothers, *toa toa*, and their cries reach up into the dark clouds and pull down the rain.

3

In the glacial north, made by ice moving across rock and grinding it into soil, rain comes from the northwest, in the summer, and falls from towering thunderheads, dark grey mountains of steam, crackling with electric tension, loosing their bright, hot current in forked rivulets down upon the wide brown earth. Winter winds turn vapor to clouds of crystals that bury furrow and field and sheath the rivers in thin blue ice for dark months on end. But water is water and spring comes and everything melts and rises and flows and changes, crystal clear icemelt into blue flood into Mississippi River dark with mud.

4

Rain blows sideways over slow blue marshes of manoomin and water lilies where little fish dart through forests of reeds. Rain falls slant across wide brown muscular reaches, hiding the far shore, the barges and docks, water like biceps, currents strong enough to take us by the unwary foot and drag our bodies from Hannibal to Ferguson, Cape Girardeau to Osceola, borne on undercurrents that make the surface boil. Rain drenches the fishers on the banks, women singing praise songs and scattering tobacco, scientists dipping samples, catching crawfish to study, drenches the paddlers in canoes downstream from Clarksdale, falls wide over the wide delta, over looping oxbows, over the fork of the Atchafalaya, over the border where cotton gives way to cane, drums on the roofs of sharecrop shacks and the cabins of lift boats chugging along channels choked with water hyacinth, falls through the leafless branches of grey ghost oaks murdered by creeping salt water at their roots.

5

There is no end to the story, but the river opens itself into this vast estuary fan, and everything lands here, a silted geography made by whatever is light enough to carry: soil, stories, industrial effluent, feathers, petals, blood, the most delicate windblown seeds of hope.

Vena Cava

The vena cava, the hollow vessel, carries spent blood to the heart to be refueled. It carries weariness into the red pumphouse of love, where oxygen crosses membrane. The renewed blood is bright red, but the tired blood isn't blue. It's the color of wine, of mulberries, of dusk. It is the waves of light passing into and reflecting back from skin and glistening layers of fat that make our veins look azure at one depth, crimson at another.

How deep will we go, one stream into another, current below current, to the bedrock and bone of what we need to know?

Adivinanza de la esperanza:[1] here is the riddle of hope. lo mío es tuyo, lo tuyo es mío, yours and mine and mine and yours, toda la sangre formando un río.

Kitípikwitana means buffalo fish in Myaamiaki-Illiniwek, not Tip-A-Canoe. Tecumseh, the Sha'wano'ki weaver of peoples and his prophet brother made warp and weft right here, at the fork of the Waapaahšiiki, which means water over white stones, the Wabash, which runs into the Good River of the Seneca, the Ohi-yo which runs into the Misi-ziibi of the Anishnaabe, the runoff of Prophetstown flowing downstream.

There is the Upper Vena Cava that flows down from the headwaters, from our eyes and necks, from the rivers of our speaking faces and gesturing hands, from the many small streams that gather in our arms, to join just below the throat, just beneath voice, mulberry, dusky, dark, tired, pouring into the right side of the churning, muscular heart. The dark blood passes through chamber upon chamber, then washes out into the great marshes of the lungs, where gasses bubble into the tiny creeks of capillaries, and the blood turns red, red, red.

---

[1]   *All Spanish quotes from Son Número 6 by Nicolás Guillén*

Wounded Knee Creek of the Lakota runs into the White River, which runs into the great Wimihsúrita, the wide Missouri. Sand Creek of the Inunina and the Tistsistas, renamed Arapaho and Cheyenne, runs into the Akansa which runs into the Misi-ziibi, the aorta.

The lower Vena Cava has tributaries and confluences just like the River. The blood pumps uphill through a thousand locks and valves, resisting gravity, climbing from our feet into our thighs, into the deep pools of the pelvis. It comes from the purple hills of the liver and the twin mounds of the kidneys. It comes from belly and womb, climbing up into the right atrium, the small sea, the tidal whirlpool of life, and flows out into the floodplains of breath.

There is the Little Minnesota and the Minnesota, the Little Bighorn and Bighorn, the Small Cardiac Vein and the Great Cardiac Vein, toda la sangre formando un río.

The Ets-pot-agie-cate runs into the Bighorn, which runs into the Mi tse a da zi which means Yellow Stone River in Hidatsa, which pours into the Missouri, which flows, dark, dusky, tired, into the Mississippi, at a place far downstream from Sand Creek and Powder River, down-stream from Crazy Horse and Red Cloud, south of the Great Sioux War, just upriver from East St. Louis in the summer of 1917, and "the bloody outrage " of fifteen dozen Black people dead, and six thousand burned out of their homes by shrieking vigilantes, while anyone who could, slipped across to St. Louis on rafts by night in the red glow of arson, and ten thousand marched in silent vigil in Harlem.

The swamp fed oxygenated red red blood flows back into the heart on the left side, into the drumbeat of the ventricle, into the main channel, the deep course, the longest river of the flesh, the Aorta, which means to raise, to lift, to carry, to carry word from upstream, lantern light from across the Ohio, carry the churning tides of the heart, carry the opening of our eyes to see in the dark, carry the legends of the liver, leaching poison, dredging sludge, healing damage, means to carry the in-spiration, the deep breath, the mouth to mouth resuscitation of hope. *Swamp fed, oxygenated, red.*

Listen. The Allegheny is the Oolikhanna, the beautiful stream of the Lenape. The Monongahela means the place of falling banks, and it springs from the ridges and hollows where the people grew corn long before foreigners called it Apalachee. The little rivers pour their hearts into the green Ohio, the meeting place of roads, where the Trail of Tears crosses the Underground Railroad, where inhale crosses exhale, toda la sangre formando un rio.

Listen. The Artibonite of Haiti once ran down River Road, a river of feet, a river of no, a river of big houses burning, and in the smoke, shadowy shapes of the possible. Met the slow spreading pools of marronage in St. Maló, seeped into the wetlands, estuaries, shallow shifting lace of bayous, lakes, spits of earth, scarred with channels, and out the wide throat of mud into the deep, out past the dead zone, the depleted waters, weary, airless, full of nitrogen, and into the roads of the living gulf, into the taste of salt, into the song of my people, *guakía yahabo, we are still here.*

Drum of my heart, push the red flood into the most delicate veins and capillaries of my extremities and let me be full. Drum of the world, in a time of extremities, fill us with red, this praise song to circulation.

Adivinanza de la esperanza, here is the riddle of hope, swamp fed, oxygenated, red as day, todos los ríos formando su sangre, all of the rivers making one blood.

A Boricua Speaks of Rivers

For Filiberto Ojeda Ríos, hero of the Puerto Rican resistance, assassinated by the FBI on September 23, 2005 while playing his trumpet, and left to bleed to death.

> *I've known rivers.*
> *My soul has grown deep like the rivers*
>
> LANGSTON HUGHES
> "A NEGRO SPEAKS OF RIVERS"

I've known rivers ancient as the ocean floor,
secret upwelling waters from the chasms of night
where the earth divides, slashed into airless depths;
known rivers without banks or beds, warm streams
where turtles travel under round dark moons,
to lay a hundred eggs in the sand, rivers
older than language, red tributaries cradling our hearts,
blue riachuelos gathered in our wrists,
tiny capillaries like creeks
buried in translucent Boricua flesh.
My spirit has learned to circle, branch, return
like the rivers.

I rise from many aquifers,
through sand and mud and stone.
I am a rainfall blown crosswise by the wind.
I came by many roads.

I washed my cooking pots in the Orinoco
when dawns were fresh and green, white winged
over the dark bank; rinsed bowls of casabe
as thunder gathered in the mountains,
and rolled brown waters down
through lakes of flooded reeds
out the great wide mouth, uncoiling like a tongue
into the blue-green sea.

I gathered red rice in the Niger delta,
filled baskets to feed a village, made cloth
from the bark of mighty trees whose leaves
reflected light of Kongo currents,
molded river clay into bowls I carved and cut
with marks of kinship and courage,
a mirror of my face.

I hunted red Iberian deer
along the Al Wadi al Kabir, lay down
in the juniper and thyme, swam with otters
and pastured sheep in its rich meadows,
watched summer's electric dance
set the pine woods ablaze,
and winter choke the canyons with snow.

I have known rivers of distant continents
washed together in the soft damp marrow of my bones.
I have grown rich with the blended mineral wealth
of their resourceful ways, whispered conspiracies
lapping at my shore. I have learned to mingle waters
and take winding paths
like rivers.

I sifted golden grains from the Toa and Manatuabón
saw the first blood of conquest stain the Loiza and Yaguez.
watched the flooded Guamani steal into the cane,
to drench and rot the calloused feet of field hands,
saw the dark arc of machetes in the smoke
chop at slavery's rope, and part the strands.

I watched grey dawn break over the East River, dark with oil,
lit by shipyard welders' flares, sewed elastic into
waistbands in airless sweatshops by the Hudson,
let my trumpet echo from the red iron walkways
of George Washington Bridge.

I have been fugitive at the cold edge of the Connecticut,
where maples burned their sharp song into the coal smoke wind,
left anonymous winter coats on the doorsteps
of my shivering neighbors,
took seven million bits of paper from a cash box on wheels
to buy them food and rebellion, arm consciences,
slip hope through a cracked window,
into a worn shoe, a threadbare pocket, to buy us time.

I have known rivers of laughter carve beauty
into the sooty canyons of heartless cities,
rivers of courage break through the sandbars of despair.
Heard music spill out into the windy street
to steal back smiles pilfered during the long dull day.
I have learned a robber's quick fingers from the flood.

I have kissed the wild and stolen waters
of my own Rio Blanco poured into a trough for soldiers,
watched children splash in the shadows of factories
full of broken thermometers, leftover poisons,
the sewage of greed, seen credits flow into far away accounts
while mercury made fingers tremble, kidneys falter,
minds grow slow.

I have known arroyos de la sierra, like Martí,
bright streams of joy tumbling from sweet cordilleras,
filling cisterns and buckets, washing the roots of yautías,
with enough left over to water my roses
in the secret garden everyone protected;

and I've seen the clear arteries of my islands drained,
piped into the pockets of strangers
who want every river on earth to bear their trademark,

MANANTIALES

the privateers of thirst, eager to sell
each quenching mouthful to people
born in the river's embrace, left licking the dust
of its dry, empty course.

I know the places where rivers branch,
the moments of choice. I know
a life can become a upwelling spring.

That morning I sang *Despierta boricua,*
defiende tus ríos while G-men wriggled
in for the kill, flat on their bellies, infrared scopes
the reporters far away, behind barricades,
and then a flashflood of bullets
through the doors of my house: I've seen that
high water coming all the years since I chose.

I died in the scarlet watershed of my lungs
and was buried in Rio Blanco under a white stone
and while the roses I once tended tangle overhead
I listen to waters gather, pool and, rise
in their secret places underground
and here in my dry and narrow bed,
I am not thirsty
I am not alone.

Baño

We enter the water singing, weeping, praying. We slide down into the healing bath. We run into the water to escape the soldiers. Singing weeping, praying. We walk into the water to find the way home. Brucha at shekhina eloteinu malkat ha olam asher kid-shatnu bi-te-vilah b'mayyin hayyim. Blessed are you, spirit of the universe, who makes us holy by embracing us in the living waters.

This is the Ohio by starlight, the creaking of oarlocks, the glittering constellation hanging like a compass over the low hills, and five hun-dred miles is a million steps, a million choices to go on, though dogs though guns though manacles though torture though fever though hunger, though thirsty always thirsty, one step after a million steps to the other side, to the safe haven, to the warm bed, to the northern lights, to the land border and the smell of maple on the wind.

This is the water of Igbo Landing, a different kind of freedom, singing the water spirits brought us here and they will carry us home, all the way, all the way home, walking into Dunbar Creek through stands of cordgrass, singing their mouths full of seawater, opening their lungs, surrendering breath to become spirits wheeling over the salt marsh, arrowing east against the wind, stolen bodies left behind like sodden logs because freedom is that precious.

This is the mouth of the Bad Axe River, the ambush, the gunboat, the desperate people running into the Mississippi, blooming into red clouds in the current, swept away downstream, children, grand-mothers, women carrying babies: the plaque says this riverbank ended a war. Clots of their blood still lodge in the roots of bayou sawgrass.

This is the Río Bravo, the border, the fierce water, the place of infrared scopes and helicopters. Two thousand miles is four million steps. Four million times of one foot before the other in Lenca, K'iche', Zapoteca, Nahua, Mixteca, Tzotzil, from one nest of guns to another, slipping down side streets, down furrows, into fields, shacks, contaminated faucets, or into jails, camps, cages, deportations, start all over again. Unless you can't. Unless your bones become white stones in the riverbed, fossils of a long war that doesn't end at the riverbank, that doesn't end.

This is the river clogged with oil, where we wade in in our thigh boots gathering drowned birds with their beseeching cries. These are the rainforest pools of Amazonas, poison rainbows shimmering over black sludge. These are the bathtubs of Flint, water as brown as any swamp, carrying all the muck of industry, infusions of rust and lead. These are the waters of St. James Parish, trapped between leaky factories dripping cancer into the groundwater, chemical plants and refineries with their burning flares day and night, pipelines ready to crack and spill, and nowhere to run.

This is the river that storm surge made of our roads, brown swirl of broken things, floating trash, shuddering people wading toward higher ground towing the sick and old and injured on rafts made of doors, their houses roofless, filled to the windowsills with flood, climbing stairs to missing upper stories trying to get dry, working from first light to last dragging debris out of the streets while faraway pundits debate whether we're human enough to be helped and a thousand bottles of water sit on a runway for a year.

Ah, but this is the mikvah of healing where we steep fistfuls of truth in living water to brew a medicine deep enough for this wounding. Blessed are you, oh breath of the universe, that embraces us in waters that still live and makes us holy again, for we were born holy, from the inner oceans of our mothers, we were born human, and we step singing into the water because freedom is that precious, we are that precious, so we submerge and rise, submerge and rise, and pull the ache of breath into our grieving lungs with all the power of our love.

This is el baño, la limpieza, sacred herbs floating on the wet skin of steaming water in a deep tub, powdered eggshell dissolved into milky blessing, the cup of spirits, the petals of roses, the handful of honey, the whisper of song.

This is the ceremony of our cells, each one a sip of water, a tiny ocean, a bubble on the bloodstream, inside our skins, beading up on our faces, trickling down our sweaty backs, dripping from our weeping eyes, offerings to the world. Then let the water in you call to the water in me. Let the water in our veins call to the water in the world, one river, one ocean, one rainstorm gathering over the dry places, ready to pour down quenching relief for the thirsty, and we are all of us thirsty.

Then as we gather ourselves in prayerful determination on the edges of endangered pools and streams, look out over the peril and grief of dead zones that should teem with life, let the water in me call to the water in you, and from the deepest waters of our beings all of us feel the pull of a single moon bright with reflected light, and we all move together, inhale, exhale, ebb tide, flow, singing shehechianu, oh what a blessing, that we have come to this time, this place, this day, knowing that the rivers are ancient, but this moment is new.

Holy Days

It's Rosh Hashana, 2016 and I am sitting on the banks of the Cannonball throwing away my burdens, giving them to this river that will carry them into the Missouri, then the Mississippi and finally into the Gulf of Mexico, and the Caribbean Sea, thinking this beautiful land is not my home, my home is thousands of miles downstream, and then the thought comes to me out of the dusk, out of the wild geese hoarsely crying overhead, out of the shadowed flight of a great blue heron, that all lands are downstream from Standing Rock.

Three hundred flags flap in the wind. Ancestors move in the air all around us, in the racing clouds, in the ever-colder wind, in the rustling of grass, they gather, drawn to our gathering. At dusk fires burn all across the valley. Horses whinny. By day, drones buzz overhead, helicopters, the whiff of chemicals. There is no difference between the lives of the people and the life of this land, this river. You cannot partition them neatly into ecology and sovereignty, green and red. They are one heartbeat standing in the path of plunder.

This ravenous pipe meant to suck oil out of sand and send it hurtling across land and water to run the engines of greed is a crime against all life everywhere. Each year these tubes, these steel intestines, rupture in hundreds of places, leaking leak fuel into fresh water and salt, to kill wildlife, sicken humans, poison whole beloved worlds. I am writing these words in Glendive, Montana, at the southern edge of the Bakken oil fields, where last year 30,000 gallons of oil spilled into the Yellowstone River that lies just across a field from where I'm camped. And now they want to risk the great Missouri.

Every land is downstream from here. In this moment, this place of prayer and drum, of human hearts and limbs knit into a web of protection, this wide sky into which our voices rise, windblown, this curve of flowing water, is a leading edge of the great decision we face.

Either we will leave the oil in the ground and change our lives to fit our planet, or allow extraction to hurtle onward at its breakneck pace, contaminating all waters, spoiling all lands, ruining the sky itself for the sake of the insatiable rich who want to be richer still, who would rather let the planet burn than share it.

I listen to the people around these fires mourn the poisoned rivers of China, Malaysia, Palestine, East Africa and West Africa, the southern reaches of the Mississippi, 48,000 gallons of diesel fuel spilled into the Wabash, the tainted Gulf coast, pools of abandoned crude oil and toxic wastes in the Ecuadorean rainforest, the destruction wrought by fish farms, super dams, oil palm plantations, indiscriminate logging. They have come here because the Lakota people of Standing Rock have made a line of hope in the path of destruction, drawn with these fires, these bodies, these prayers and songs and life stories, this grief and solemn joy.

Downstream from this silver streak of water at dusk lies the wide, blue, salty Gulf, its edges laced with oil channels, pipes, rigs, lies the southern end of the pipeline and the bayou camp of L'eau est la vie with its banners, tents, fire, song. There is the sacred Gualcarque River of the Lenca, for which Berta Cáceres died. There are the thousands of branching streams of Amazonas, and the lacy web of water that wraps our planet, water moving through all lands, feeding all oceans, becoming all rain.

These are the high holy days of the world. Life and death have I set before you, we pray on Yom Kippur morning, therefore choose life, so that you and your children may live. The nations gathered under the wide Dakota sky, and all of us who have come to stand with them, are choosing life, stubborn, persistent, beautiful life. Everyone on earth lives downstream from that choice.

When I Write River

The river I write is not only this actual vein of rainwater and clay, this great branching waterway wrapped like a web of artery through and around the heart of a continent.

Although I am singing to a single blackened pine needle, one grain of sandstone, one teaspoon of spring rain over the Alleghenies as they dance their way toward the blue gulf, chanting that exact whirling trajectory of flood and flow we call Mississippi River, the river I write is flow itself.

I am writing what pushes forward, what cuts across continents, grinds into mountains, meanders across plains meeting itself, only to cut off oxbows of the useless sidetrack, seeks its course, feels its way toward the right road, making canyons and swamps and rapids only because this is how water travels through land, because water trusts its destination and reaches for it across whatever is in its way, including cities and levees and mistakes made in ignorance or hubris or small minded acts of grabbing for the moment to possess a fistful of current.

When I say river, I mean the long persistent desire.

When I say river, I mean faithful to a distant freedom as far beyond what we are living now as the wide-open sea is beyond each winding creek feeding the stream, but which calls us to find our true level, to overspill the brimming puddles and pools of our confinement and reach and keep reaching, because to reach again and again is what makes current.

When I say river, I mean reach.

TWO

Mar

Yocahu Bagua Maurocoti ::
Yemaya :: Asherat

I Came Here from the South

Although I drove the edge of the River from north to south, following the current downstream, stopping to dip my fingers, to offer tobacco and sing, I came here from the south. I came to the river's mouth from the sea. I came from the nineteenth degree. This is what I know: salt, conch shell, sand, hurricane, phytoplankton, hermit crab, West Indian Top Shell, phosphorescence, coral reefs, cayos. I am archipelagan.

The warm sea-river, the gulf-stream, snakes through the sparkling cool of the Atlantic into Bagua, the small sea, hot moving through cold, spawning vortices of water and wind. We ride from island to island in its warm embrace, gather up the silt laden river and move along the curve of the continent *singing guakía yahabo, we are still here.*

My people came out of the mouths of many rivers into the salt spray and the heaving swells, crossing oceans to reach islands. They came from the *Orinoco, Toa, Niger, Guadalquivir, Dnieper, Hudson.*

1

Rain falls hard on the Parima mountains and divides into two, like hair parted by the comb. The eastern water flows down to the Branco, then the Negro and into the Amazon which has a thousand other, older names. The western water flows into the Orinoco. In the heights where the waters fall and part and flow, the Yanomami live in the path of iron and oil, and are dying of measles.

When Deminan and his brothers fled the wrath of their grandfather, the father of mother earth, it was from the mouth of the Orinoco river that they paddled their canoa, hurtling out into the ocean, out of sight of land, and followed the island chain north and west until they came to Boriken. Orinoco is the first home of the Boricuas.

The second home of the Boricuas has a spine of serpentine where clouds gather and rainwater flows downhill to all the coasts. My people built their houses along the mother river they called Toa until my other peoples came, some free, some in chains, and named it Plata, which means silver. Along the Toa, my first people grew yuca, batata, yautía, papaya, caught fish and hunted birds, and along the Plata, my second peoples planted rice and tobacco, raised cattle, worked the cane, and together they all grew hungry, and rode to the harbors where the steamships puffed and tugged at their anchors, and steamed away north.

2

Ay, vientos alisios, trade winds of the Atlantic, what have you brought us from the teeming docks and counting houses of Sevilla, out of the mouth of the Gualadquivir, al-wadi-al-kabir, river of the great valley? From this inland river port my Cantabrian ancestors, my Iberian people, soldiers and shepherds, captains and working people, hoping for rescue, hoping for riches beyond dreaming, set sail for the place they called America on a wind of desperate hope and greed, and fell like a rain of stones on the green lands, on the island people.

At Los Frailes Mine, on the banks of the Guadalquiver, a dam ruptures and 180 million cubic feet of tailings, the gristle from the meat, what the mines spit out after zinc and silver are torn from the ground, spill into the river and the stain flows out to sea. Their faint metallic tang washes shells on the beaches of Ceiba where the solvents and sludges of military occupation seep through groundwater to meet it, in among the mangrove roots.

3

Ger-n-ger, river of rivers. The great bend of the Niger River, seen from among the stars, is a green arc through the brown lands of the Sahel, and then there is a tangle of green threads, braiding and unraveling, the inland delta between Segou and Timbuktu. It is a river born among rocks, so it runs clear instead of dragging the mud of a continent. It is the water road through Saharan drylands, the trade road of Mali and Gao, and from its banks and marshes my people were dragged in blood, bound and stacked, forced onto ships and blown away across the deathly middle passage, to be cast, those still living, seasick, heartsick, exhausted, onto the docks, onto the sandy inlets, onto the sweet earth of our islands.

The Niger is home to two hundred and fifty species of freshwater fish and the dying West African Manatee whose low song laps against the reefs and inlets where only six hundred Antillean manatees survive, singing their sorrow back across the sea.

4

There are nine rapids on the Dnieper River as it runs past the glowing perils of Chernobyl, Prydniprovsky, the slow seepage of radioactive wastes. There are nine rapids on the river that runs past Kremenchuk where a rabbi's wife defied the law of a male God and walked away from the synagogue to become my great great great grandmother. There nine rapids and there are aging dams poised to fail, ready to flood the same earth that armies overran, still full of mass graves, to flood the dark, rich earth that could feed a continent. From the port city of Odessa, my people climbed onto ships that left behind the smoke of the pogroms shrouding the earth they loved. They didn't go to the warm islands. They went to Manahatta, crowded city of tenements and skyscrapers. They went to a place they were told was golden. They went to the Hudson, the *muh-he-kun-ne-tuk*, the river that flows both ways, one more shipload of settlers in the land of the Lenape.

5

Orinoco, Toa, Niger, Guadalquivir, Dnieper all gathered in the veins of my parents, as their people gathered in garment sweatshops and janitors' closets, surrounded by the Hudson, a drowned river, its mouth buried out at sea, a salty estuary that surges and ebbs all the way up to Troy, surrounded by the riptides and currents of a basin deeper than the coastal ocean, a swirling, rocking pool of people and waters mingling, of sooty tugboats and barges, spangled bridges, dirty buildings, tired people who still make music, the scent of their spices rising from crowded tenement kitchens, wafted between buildings.

When they left, the Hudson it was not by boat. They went winging southeast on rivers of air, and one afternoon in the dry season, as the coffee berries ripened toward harvest, water rushed from my mother's body and bore me into the light of a Caribbean day, among the rain-fed springs and creeks of the highlands, in a tiny wooden hospital at the edge of the Río Castañer.

6

I have walked the riverbanks of the north, touched their cold waters, lived by the Chicago, the Gale and the Pemigewasset, the Cambridge and the Mystic, and the great Sacramento with its inland delta, refuge, like all swamps, of the dispossessed. But wherever I go, whatever arrow of the compass rose I face, I come from the sun, from the circling storm, from the sea salt wind. When I face into North America, I come from the south.

The River is Older Than the Sea

The river was first. The river is older than the islands. Older than the little sea. Its stones were formed in an age of lava and ice, at the same time that algae invented photosynthesis. In the bedrock are carved the memoirs of oxygen, from its first fermenting out of the blue-green scum of shallow waters, until their toxic exhalations extinguished almost everything, and forced evolution down a new path.

Oxygen seized the iron out of that mineral broth and bound it into thick layers of ore, and so the Mesabi and the Gunflint and the Cuyuna were made, inscribed into rock by the gases of the air.

Basalt and sediment rose up and were ground down. Fine quartz sands drifted into future prairie soils, bubbles of gas percolated up through lava and scattered agates in the earth. It was still a thousand million years ago.

For millions upon millions of years the land was tropical, washed by rising and falling seas, a single continent, one gigantic slab of rock and burgeoning life. Then Amazonia and Benin broke their embrace, Africa parted from America, Pangaea broke apart and slowly spun out and away in great floating chunks. Life split into new branches, into separate dances of evolution, and sometime in those years of spinning and floating, as the small sea lurch eastward, the islands marking its fiery rim broke the surface of the salt water and became an archipelago.

The islands were born after the archaeopteryx and before cranes, after the bees and crocodiles, but before sharks. There had already been a billion generations of beetles, but mammals were new. Winged insects flitted among ferns. The world had experienced the first kangaroos.

Waves of mass extinctions passed over the planet, across island and river worlds alike, wiping the slate almost clean, burying bones and shells and carapaces, leaving just enough for resurgence, beginning again, which life does from the smallest smear of potential on the underside of circumstance.

When the dinosaurs fell into their swamps and died, and mammals crept out of the underbrush to run around in the wide and newly emptied world, to eat insects and leaves and multiply, the islands already frothed in their girdles of sea foam, while the great plain of the river continued its dance of water and fire, glowing lava and weedy wetlands.

After a while, monkeys rode their rafts of matted brush across the gap and began to differ, American monkeys from African monkeys, Orinoco from Niger. After another long while the air was filled with the wings of pigeons in immense flocks that made their own wind when they rose, like the thunder of a thousand waterfalls. The forests were lit by the plumage of parrot and macaw, shimmering green, blazing red, blues that burned, bursts of yellow, shocking orange, sleek purple. People were not yet born, but flocks had erupted into the sky.

Almost the other day, when our ancestors had only just stood up on their back legs and begun wobbling toward their humanity, when they were already squatting and knocking chips off flint, one rock against another to make the first tools, ice came and scoured the river world.

It was a dance of back and forth, advance and retreat, huge grinding lobes of blue-white pushing south, carving out the softer stone, and then crawling north again, leaving tooth marks and potholes, leaving coarse moraines, and scattering fine-ground limestone and granite and shale.

Along the rotting rim of the ice, chains of lakes formed. Streams of melt water tunneled through and left winding ridges of sand, inverted riverbeds snaking across the flatlands. Wind came, sculpting what the ice left behind, lifting the finest grade of soil and flinging it downwind, across future grasslands.

Immense ancient rivers flowed out of mammoth glacial lakes, leaving behind terraces and valleys that dwarf the little rivers of today, the Minnesota and the St. Croix and even the Mississippi, just a rivulet compared to its great mothers. It was from the tangle of lakes and ice-born streams that the river was made, under the rosy fires of the northern night sky.

And out of the clash of earth's crust, of plates moving and grinding, spewing lava as they go, the basin of what would become Caribbean, the belly bowl of new molten rock and seabed, pushed eastward toward the rising sun, through the still widening waist of America, diving beneath the Atlantic, releasing an arc of fire rising through water, a string of volcanic beads, cone shaped islands that were its edge, steaming and crackling as water and fire pressed against each other, and ocean floor tipped sideways into topography.

Molten rock cooled into the spines of archipelagos. Limestone accepted rain into its hollows and dissolved. In the rich crumbling stones, birds dropped seeds, seeds made earth, earth ate rock, sand drifted against the edges of mountains, buds of coral latched onto submerged limestone and began to build their infinite cities in the turquoise water. Land in water, water in land, river and islands came into being, rich in story, long before the first human foot made its first scuffed mark on the ground.

In the beginning

In the beginning we were made of mud, of the bowels of the sun, of a flake of flint. We climbed out of turtle eggs and holes in the earth, fell from the sky and rode to earth on the backs of geese. We were the younger siblings of the taro root, stubby kernels of the corn mothers, tiny figures made of clay and breathed into life. We are made of what everything is made of, only later.

Salt

Some say it was the rivers that gave the seas their salt. That rain pounded on volcanic rock for thousands of years, century after century of rain passing through air, through carbon dioxide, and becoming ever so slightly sour, just enough to dissolve the surfaces of stone. As rain fell, age after age, eroding the rock away, molecules danced, attached, were elemental. Washing into the sea, the creatures of the deep water sucked up their favorite minerals to make bones and shells and soft innards, and left the rest behind. In the churning oceanic waters, sodium with its positive charge and chloride with its negative, consummated their irresistible attraction and made salt. And now the salt in the seas would make a crust around the world five hundred feet deep.

It cakes at the rims of tide pools that pock the ancient fossil reefs of Boriken, smoky grey, prickling with sea urchins, limpets, blue crabs skittering sideways into crevices. It lies in dry, shining layers among fossil fish, in deserts whose flat expanses were once inland seas. It sparkles in the flanks of the highest mountains on earth, lifted out of the ocean to the edges of our planetary air.

Without it our hearts would shudder and gasp. Our nerves wouldn't fire. We could not make enough blood. Our muscles couldn't move. Our cells would get waterlogged. Animals walk their narrow tracks for miles to lick it from its shallow places, where seams of it break the surface of the ground.

Standing barefoot in the sand, right at the farthest reach of the waves, the briny sea washes over my ankles, recedes and leaves my feet shining with salty wetness. We are made of what the oceans are made of, salt in our veins, salt in our synapses, salt in the words we speak.

Phytoplankton

They float on the skin of the sea, drifting, beautiful, microscopic sparks of life, tiny geometric jewels, turning sunlight and seawater into themselves, and making oxygen for all beings that breathe. They are far more numerous and diverse than the birds of the rainforests or the flowers of the fields. They make more air than all the forests of the world combined. They glitter and float and soak in minerals and light through the luminous walls of their single cell bodies.

They are the first food, the oldest meal, the deepest nourishment, and everything in the ocean lives from their tiny bodies as unnumbered as stars, millions of sparks in a teaspoon of seawater. They are the ancestors of our ancestors. They take carbon into themselves, pulling from the sky, the winds, decaying leaves washed out to sea. They bind it into their beautiful crystalline bodies, and when they die, they fall slowly down, down, down through layers of midnight spangled with phosphorescent fish, onto the seabed, and ancient beings that they are, there are great drifts of their bodies, whole sedimentary mountains of their dead, pressed between the oceanic weight of water and heat of the earth's fiery core. Their minute, sparkling selves dissolve into paleo-blackstrap, the simmered syrup of time, fossils of light, the deepest arterial pools, charged with sunlight, buried under stone. Slowly, slowly they become the substance we call petroleum, rock oil, black gold, the wealth and peril of the world.

Bioluminescence

There are only five shining bays in the world that sparkle with this blue-green light, and three of them are in Borikén. Delicate galaxies of tiny flashing beings speak in light each time the waters are stirred, make bright ruffles to the waves, shining wakes to boats, drip cool fire from our oars.

It's only at night that they flash and twinkle, startling predators, rolling with the waves. A blue flash of lasting only a tenth of a second, drifts of living stars winking in and out of sight, constellations that only appear when the waters are troubled. But the trouble of these waters is too big for them. They are early warning beacons, delicate phytoplankton going dark as the water heats, as storms batter the bays and inlets where they glow, as diesel and run off drain from the land and trickle into the waves.

Tiny mothers of air, the phytoplankton, those that spark light and those that do not, are falling off the cliff of extinction. What then will we breath in the starless dark?

Coral

Tonight the moon is full. In the warm, salty waters of the Caribbean, coral polyps release clouds of spawn at sunset, making masses of larvae, while those embedded in their underwater cities joyfully divide and multiply. Out of their own bodies calcium seeps and hardens, building, a centimeter a year, great banks and reefs where shellfish and snails anchor, and bright fish dart and hover, grazing on algae, big fish eating little fish, eels and starfish, sea urchins and sponges finding homes in the nooks and crannies of living stone.

Thousands of miles away, the Great Barrier Reef has been two million years in the building, but in the last three years, half of its vast expanse has died, as the algae it feeds on perishes from heatstroke in the simmering sea.

With algae, it's all about balance. Too little and the coral starves. Too much and it suffocates.

Here in the Caribbean the sea is warming, but it's mostly other threats that give the reefs another twenty years to live. For the algae to stay in line, feed and not smother the polyps, the reefs need parrotfish and sea urchins to graze their surfaces like living lawnmowers.

Parrotfish have a thousand crystalline teeth, harder than metal, set in rows. When one set wears down, the next is ready to step up. They eat the overwhelming algae and munch on the dead structures of coral, pooping out hundreds of pounds of fine white sand. Dazzling in their brilliant scales of blue and gold, orange, red and green, they change sexes across their lifetimes, wrap themselves in membranes to sleep, and mate in groups of females sharing a male. They tend the reef and keep it healthy.

But there is runoff from golf courses and hotel parking lots, there is fuel leaked from the engines of boats, and there are the hooks and nets of humans, taking far too many of the bright, beaked fishes, so the algae go wild, coating the polyps in sludge and choking them, leaving bleached, silent reefs, lifeless as ancient ruined cities abandoned to the tides, mass graves to be excavated by our descendants, evidence of the crimes of ecocide.

The waters will go on warming, but there is still time to protect the parrotfishes, time to stop fishing them out of their reefy homelands, time to stop leaching poison into their sea so they can graze the reefs, and keep the living cities of the corals alight with life, at least for now

When I Write Archipelago

The islands I write are more than water-bordered lands. They are mountain ranges rooted in saltwater, bones of connection visible only from the air. I sing the islands one by one, their surf ruffled edges, their sloping beaches and plunging cliffs, the rings of reef-sheltered bays, the wild oceanic swells grinding stone to sand. I sing the dry ones where every drop of rain must be cupped, and the wet ones, of palm and fern, moss and orchid, the volcanic peaks and the flat slabs of uplifted limestone, the twelve-hundred-mile arc of Cuba and the tiny specks of scattered cays. The archipelago I write rises from the bones of the planet and is interrupted by reflected light, rises above the skin of water and continues far below. When I say archipelago, I mean continuity and uniqueness, how we are singular and insepa-rable, boundaries of rough surf, rough passages between our coasts, and one shared bedrock, one tectonic gap, one subduction of vast planetary stone that raises us. When I say Caribbean, say Antilles, say West Indies, say Windward and Leeward, Greater and Lesser, I mean that we cry out in different languages as the same storm sweeps our houses into the sea. When I say islands I mean the endless dance of kinship, pulling us together and apart like tide. When I say islands, I mean we.

THREE

Wind

Guabancex :: Oyá :: Ruach

Sacrifice Zones

Levee Camp, New Orleans, 1912: The river is in dangerous flood, lapping at the tops of the levees, so the white men who carry guns make the black men who work their fields haul bags of sand onto the narrow ridge of the levee. This is the slavery of the convict gangs, men captured and jailed and brought to the edge of the flood to break their backs as they do in the cane. It's the servitude of sharecropping, the bondages of debt. But the water keeps rising, and there are gracious homes at risk behind the wall. The flood drags at their knees as they throw the sacks of sand down and go back for another and another, but the engineers are running out of sand, so they make an offering to the god they worship, a sacrifice. They give the men one choice: lie down on the levee and make your own bodies into a wall to protect our property or be shot where you stand. So they lie down for the slim chance, as water fills their eyes, their mouths, their lungs, drags them from the high ground and away into the thick, brown, angry current as it fights its way out of the levees, silting the wide land beyond with mud and bone.

When I think of my own country left starving, thirsty, hot, when I think of water withheld and un-repaired wires dangling and no light, no fans, food gathering dust in locked warehouses, roads left buried in mud; when I think of un-repaired wires scattering sparks across dry California acres in a high wind, whole towns burned alive, vast plumes of suffocation, all to cut costs for the corporation that is demanding we the people dig them out of the pit of ash they made, I struggle to breathe. I feel the flood rising toward my mouth and I know they are laying all our lives on the levees, building them higher with our bodies, making reefs from our bones to hold back the sea, hoping to stay dry, cool, sated, behind this wall of sacrifice.

Three thousand acres an hour, the Amazonian lungs of the world are bulldozed into raw scar, into palm oil plantations, into miles and miles of soybeans fed to cattle and pigs in faraway feedlots for the sake of bacon burgers. Three thousand acres an hour fall, crushed, splintered, burned into the sky, and the people of the rainforest are crushed, splintered, burned, plowed under. The bad breath of oil plunder, the gunky pools of residue lie still, reflecting nothing. The new president of Brazil says if the people get in the way of industry they will cease to exist. He burns the treaties, and already men with weapons are burning villages.

Meanwhile the people of coasts, of the estuaries, of the salt marshes, of the atolls lie down at night and listen to the tide line creeping toward them, almost to their mouths. All the world is a sacrifice zone, dotted with enclaves for the rich, who imagine their money will keep the heat away. They abandon their beachfront properties and take over the once scorned ridges of the poor, while down in the bayou, St. James Parish is a village in a ring of poison factories, tethered in place as the giant, devouring snake draws near, belly filled with fuel for the flares that burn all night, for the leaky tanks, for the engines of commerce.

All the world is a sacrifice zone and among the first to go are most of the freshwater fish and frogs of the Americas, seventy percent of all vertebrates, the clouds of insects without whom nothing blooms. One by one the men lying on the levees blink out, eyes going dark in the muddy water, and one by one species flicker out while gazillionaires keep their eyes fixed on rising balance sheets, ignoring the oil slick waters around their ankles. They would rather destroy the world than share it.

So here I am, an organism long sickened by pesticides and the vio- lence of extraction, unable to go with the chanting multitudes and drag my body to the doors of banks, to make part of a reef of torsos shouting in their lobbies, and so I split my tongue into a thousand voices and keep writing, sending out words like rope: hold here, hold tight, don't roll off the levee, don't drown, synchronize our heartbeats with song, refuse to be the burnt offering on the blood soaked stone.

Reaching Climax

*"We are what we do, and most of all what we do
change what we are."*

EDUARDO GALEANO FROM
"IN DEFENSE OF THE WORD"

Successional species climb one over the other, each one changing the
ecosystem, creating the conditions that will make its own life impos-
sible, making of its own extinction a new and improved possibility
for the next in line. Grass and helechos make soil for majaguas and
cottonwoods and grow smaller in their shade. Juniper and yagrumo,
pomarrosa and pine shoot up in the gaps, casting deeper shadows
over the soil the grasses made, and the sun-hungry first comers
wither in the dim light. But in those shadows the seedlings of hard-
woods, of maple and oak, higüerillo and capá begin their long, slow
climb toward the canopy.

We think we are the apex of evolution, but maybe we are grass. Maybe
there are a host of other species waiting their turn in the wings, seeds
and eggs and embryos tucked into the odd corners of evolution, ready
to spring forth and multiply if we make a habitat too hot to hold us.
Maybe we will only manage to prepare the way. To pass into the bed
of nutrients on which the heirs to our actions thrive.

There is a moment in the life of a forest that is called climax, when
waves of plants, each riding the transformations of another, reach
balance: leaves rot into humus and hold water, earthworms break up
clumps and let in air, tangles of rootlets keep the soil from wander-
ing. There is shade, and in the gaps of fallen trunks there is sun for
fast-growing seedlings to begin again. Berries and blossoms draw

insects and birds; seeds fall, pollen migrates. Rodents and fungi, foxes and bats, pond scum and lichen, monkeys and moss, and trees dense enough to call the rain, together become a world, each feeding each, no one replacing another except in the slow inching along of evolution, mutating and shifting into a new finch's wing, a different shade of petalled pink.

What if we chose to give each other shade, to deposit our rotting leftovers by the stems and trunks of each other's lives, to join all our filaments and tie the clay and humus and mineral grains into a great heaping table of food for all of us? What if we became the foresters of the forests we are? What if, for the first time in evolution, one small and dangerous species decided not to become successors to half the great diversity of life? The algae who poisoned the earth with oxygen and killed almost everything alive had no choice. What if we choose?

What if we do not extinguish the others? What if we do not make a bed we must lie in and cannot survive? What if we do not eat up our ecosystem and go extinct, taking a galaxy of small worlds with us? What if we coax the great woods back so we can breathe each other, our exhalation their intake, their exhaled oxygen kissing our lungs? What if we live by the story that each kind of life is essential and we do not succeed them. What if we all reach climax together?

Coaybay

Before I speak, I make a circle of chalk, a circle of efun made from these words, cáscara, cascarilla, caracaracol, white powder of egg-shell, white powder of salt, a circle of protection, a roundhouse for the dead I am calling.

The Taino land of the dead is Coaybay, and it lies on the far side of an island to the west, on the dark side of the moon. The opías, the spirits, live there, and fly out at night on bats' wings to eat guavas and seduce the living with their velvet touch. But when the sun rises, they return to the great cave in Coaybay, into the womb of the earth. I watch for them, in this twilight. Come help us, I cry, or there will be no rebirth. I conjure hope from a bat's shadow at dusk.

What if Coaybay is a great kilombo of the dead, a maroon world of ancestors who have been waiting for the right moment to be reborn? What if the ancestors are tired of feasting on guavas while their children's children's children are forced to lie down under the guns of greed and watch the waters rise toward our mouths?

What if they have decided that now is the time to return, generation upon generation of ancestors in a torrential downpour of help. What if the rivers and seas return them to us, not terrifying Hollywood zombies devouring human flesh—the ones who eat us wear suits and do not smell of earth. What if they are returning in the strength of our heartbeats, in the redness of our blood, as seeds? What if the dead rise up out of Dunbar Creek to bring us medicine? What if they come from ten thousand corners of the world, from the dark side of the moon, from the deeps of the Middle Passage and the length of the Trail of Tears, from their place among the constellations and from the ground below our shoes, only to lay the silt of their lifetimes at our feet, and make a place for us to walk?

Paleochannels

Underneath the works of civil engineers lie the veins of ancient rivers, ancestral courses that wind in gravel and sand far below the surfaces where we walk and build and plant and drive. Earth has buried them, but in the gaps between their coarse pebbles, maroon water seeps under the levees and fans out across the land the river built, spring flood after spring flood, dropping its thick, brown silt at last, laying down soil.

The makers of cities and shipping lanes have edged what they think of as the main channel with concrete levees, building them ever higher, forcing the River to flow where they want it, even as it yearns toward the Atchafalaya. The River has always changed its course, sought new pathways, made six great tongues of land out into the Gulf, but constrained by walls, it can no longer enrich the earth and the land washes away, sinks down into itself, disappears.

But the paleochannels, the secret rivers, fan out in every direction beneath the constructed banks, take their tithe of the river's flow, and reach out as dozens of watery fingers toward the coast, slipping away from the city into the swamps. This isn't a trickle, a damp patch, a seeping. The water that runs away is as much as the whole Hudson river pours past the towers of Manhattan. Water finds its way.

Out in the wetlands at the Gulf's edge, more than an acre an hour washes away, collapses and sinks, drowning trees and fields, houses and roads, a landscape hungry for the stroking hands of the river's flood that lays earth over earth, building up soil, making land, making islands, banks, solid ground. Meanwhile the river growls in its narrowed bed and experts bicker.

A day will come when the levees will break apart and the clenched fist of the river will open into a generous hand, remaking what was stolen, but for now, I chant to the paleochannels, rebellious water running along the old paths, and river roots taking hold of the land that the river made.

Insect Afterlife

I dream I am walking through meadows filled with the insect dead, the afterlife of Monarch butterflies and tiny midges, of ants and mayflies and beetles, with ghost crickets chirping in the underbrush.

In the rainforested mountains of Yucahú, abode of my ancestral gods, the leaf mold underfoot no longer crawls with tiny legs and the hum of tiny wings falls silent, while ghost lizards wander, hungry. This richly woven world of dripping fern, waterfall, mountain palms, towering mahogany exists in a narrow band of warm, and as it edges toward hot, the insects die of heatstroke, and the lizards and frogs and birds that feed on them die of hunger.

In the insect afterlife the air is full of their humming, the vibration of their movements making the air quiver. There is a great source of power in the shimmer of their dances, and they say, use it, try if you can turn the tide. I throw wide my arms and am filled with the vast flickering multitude. I become a filament alight with the sparks of their passing. Now, everywhere I go, I feel them around me, the brush of moth wings, clouds of tiny forms in every sunbeam.

I remember fireflies in the cool evenings of the cordillera, and a deaf-ening chorus of coquis and crickets. Come to me, I call to them. Give me your light. Give me your song. Give me multitudes. Give me strength, I say, and I will try.

Unobituary

Writing about hurricane devastated Puerto Rico from the midst of a Northern California in flames, it couldn't be any clearer that climate catastrophe has arrived. Planetary warming erupts into hurricanes and wildfires, more often, more widespread, more deadly than before, and the apocalyptic landscapes of San Juan and Santa Rosa look eerily the same.

On lockdown in my tiny eco-home to avoid the choking pall of smoke outside, I scour the internet for raw materials. Hidden in the crevices of the info highway are the people who help each other stay alive, who plant, rebuild, invent, imagine, make ceremony and marches, coops and strikes, while the headlines are dominated by robber barons eagerly destroying every obstacle to plunder, unable to stop craving the total ownership of everything.

Reading the articles on Puerto Rico that people post on Facebook is like having a subscription to a horror movie channel. It's hard to remember all the human and material support that's flowing toward exhausted organizers, who are trying to lay the groundwork for a shift that goes beyond immediate survival, into an alternative future of climate resilience, food security and sovereignty.

I began this day filled with images of smallpox blankets being handed out by FEMA, of the locked warehouses of the Irish potato famine, full of grain that could have saved a million lives, of dystopian nightmares in which the desperate must fill out mountains of forms, and age and die while standing in line.

But then I started hearing the voice of Nuyorican poet Pedro Pietri, remembered him in his flapping black clothes, leaning into the microphone, ferociously detailing the everyday deadliness of our condition, and how we chanted along with him: *all died yesterday today and will die again tomorrow.*

Puerto Rican Obituary, written by Pietri in 1969, is a kind of national anthem of the Puerto Rican diaspora, its names and refrains as familiar to us as any classic song, and many poets, following the communal musical traditions of our people, have sampled and riffed off it to write our own declarations. There are enough of them to fill a book. This one is mine.

Nothing in this poem is made up. People did stand in line all day in the hope of food and water, only to be handed mops and cleaning supplies. People going to emergency centers really were given a phone number to call in a country without working phones. And I am as angry as Pietri. Sparks flying from my hair, my fingers, my tongue.

So, this isn't a fancy piece of literature. It isn't about the torn beauty of my native land. It doesn't say everything. In fact, it says hardly anything.

I didn't polish it into perfection, I hacked it out of the headlines, and threw it like a stone.

All died yesterday today
And will die again tomorrow.

JUAN MIGUEL MILAGROS OLGA MANUEL

All died
    waiting in line for a sip of water, waiting for arroz con algo
    from benevolent rulers who gave them
    mops and bottles of floor soap,
    boxes of hot sauce, mayo, and ketchup,
    as if to say we don't recognize you without a mop in your hands,
    we gave you the condiments, now scrub for your bread.

All died
> In a rain of insults and paper towels
> And he said they were lazy, shiftless, nasty,
> that their hunger was a hole in his pocket
> that it wasn't a real disaster like the ones real people have
> but he could spare them a trophy.

All died
> Filling out forms that are the only dry things in their lives
> Applying in triplicate for a piece of blue plastic
> to hang over the ruins. where they lie
> parched and sweating, holding each other close,
> Proving their eligibility to receive a snack sized bag of Cheetohs
> as a first installment on their malnutrition.

All died
> Holding pieces of paper with the cruel joke of a phone number
> In a country without phones.

All died
> Waiting for the promises of colonial citizenship
> To knock on their doors shouting *Mira! Mira!*
> *It was all a bad dream, we brought you real meals,*
> *A million water trucks, all the packages*
> *your familia worked so hard to send you,*
> And then they all died again from waking up empty handed
> day after day after day.

We interrupt this obituary to inform you
> that Juan went door to door on the thirteenth floor
> Checking on old people stranded by dead elevators
> then he skipped the ATM line
> went to straight to the bank manager
> and asked for a refund of the last 119 years.

> Miguel shared his generator with the neighbors
> so one of them could charge up the machine
> that keeps his little girl breathing, and one of them
> could keep her insulin cold, and one of them

could print out her manifesto on the power of people and sunlight
before the diesel ran out.

Milagros organized a brigade to open a mountain road
    Collect broken green bananas from broken hillsides
    and build an outdoor kitchen to boil them in a trickle of creek
    water
    so her hungry neighbors could gather around one pot
    for a bite of guineo and a feast of each other.

Olga has been climbing trees and mountainsides
    Hunting for a signal, tossing her text message telegrams
    into the cyber-sea:

    There is no food, no light, no water, no gas,
    No FEMA, no politicos, no doctors, no help.
    We've grown thin as stalks of cane,
    And we're hoarse from shouting your names.
    So, if you hear me, she says, then listen up, world:

There is no good news story here but us, and we are learning
    how to *be* each other's power lines. How to be
    houses and bread. Our thirst writes
    banners into the air. We are becoming seed
    of a different harvest.

Juan Miguel Milagros Olga Manuel
    All fought back yesterday today
    And we will fight again tomorrow.
    This is the voice of Borikén
    broadcasting from el derrumbe.
    We'll be back in the morning. Pass it on.

Sediment of Spirit

Silt is also the layered sediment of spirit, small acts of bravery like particles of transported soil, piling up into great sandbars that can change the course of history. Nothing is lost, no act of bravery, no moment of solidarity, no spark of integrity is ever lost or wasted.

Listen. There are moraines and reefs of courage, deposits in the spiritual landscape of the continent and archipelago. Black Hawk is not gone though he is buried. Nanny of the Maroons is not gone, though her body has decomposed into tropical soil. She is lodged in our memories like a great blue Jamaican mountain although she is dead, and the great risen masses of Haiti are dead, and thousands of bones lie cached along the trails of weeping out of the old southeast into Oklahoma. Though Tecumseh's fallen kin fill the fields of Indiana like drifts of stones, are lodged in narrow graves of Kansas dust, the Shawnee star rises above us all, still summoning us to braid our destinies, still calling us to come. Nothing is wasted. Love the courage it makes erode most slowly of all human parts. When the soft tissue of fear and sadness has sunk back into soil, these ribs remain, ridges of high ground from which to watch the horizon tilt toward day.

If we could see with the eyes of spirit, a new topography would emerge, a different terrain of rivers, mountains, plains fed by intention, choice, right action, a ground that persists within us and on which we stand. A great poet, a volcanic range of a poet from the fiery middle of America, spoke these words that map the balance of power. He said "all together they have more death than we, but all together we have more life than they", and so it is. Life is with the life-givers, whether they live or die, death abides with the givers of death, and life is stronger. Life given stays in the ecosystem of our souls, and continues to give life.

The soil of Prophetstown is hallowed by the great hope that kindled there, made holy by the will to be people together in a wilderness of contempt. Sanctified by the refusal to commit torture. The steep slopes of maroon towns are rich with the potent magic of escape, thickets of defiance, the cultivation of freedom. The small reef ridden body of Vieques blazes on this map, meeting place of thousands of threads of light. Black Hawk rises above the fields and prairies, a great dark hill of resolution. Fannie Lou Hamer is a constant, radiant ember in the soil of Mississippi, and Carlota Lucumí rises again and again in cane of Matanzas. No-one is finally defeated.

Whatever the despair of Black Hawk, chained through the bitter winter of 1832, paraded through settler cities and given into the custody of Keokuk, whom he despised for his cooperation, for signing away Sauk land and life, his choices stand above the flat plane of discouragement, an outcropping of fuel, ready to light us now; as Meridel's Girl lights her corner of the pitch black Minnesota winter night, and Zora's Janie rises from the poisoned swamps of the south in a breath of honey and pear blossom. Nanny laughs in the wind smelling of allspice and gunpowder, and Anacaona steps out of the flames like a ripe moon. Filiberto's trumpet notes rise in our chests and we breathe deeper, stronger, grow more roses to give to our people shivering in the cold Hartford night. Every brave note of jazz ever sung in a New Orleans bordello, every cornet riff played to lift their spirits above the lust of the johns by women determined to thrive, still lifts us. Satchmo's blast of brass and el songoro cosongo son de Guillén jam together across salt water full of sharks, and nothing is lost. Nothing is ever lost.

Wild Wind

Sometimes a wild wind blows through my body and makes me spin and fall like the trees of my homeland, a river of air that overflows its bounds, electric, crackling, tearing limb from limb, far beyond the cadence of breath.

Who knows where the seeds of these storms lie, in what tropic of my synapses hot meets cold, pressure falls, the spin begins.

At the edge of the Sahara, hot dry air collides with the cool moist breath of forests and the shock of it makes wind, makes the African Easterly Jet, which wobbles, unstable, in the high atmosphere, making waves, making troughs that roll out to sea. If these waves of air are wet enough, if they have instability and lift, they become thunderstorms, clusters of them, and some of them encircle a clearing of calm at their hearts and begin to spin. Every few days waves form near Cape Verde, but it's not til the season is right, ripe, ready that wave after wave spawns storm, and hurricanes spin out on their curved tracks toward the islands and shores of the North American subtropic.

I must have not slept, not eaten, been distraught, had deadlines, been loaded up with the dampness of worry, must have had many different wobbles begin to move together, begin to turn and circle back on themselves, for a cyclone to break loose and words flow away from me, so that I fly up out of the top of my head, I fall and fall and fall and wake up hours later on the floor, broken, sometimes broken, sometimes shattered like a tree of my homeland, rib fractured, bruises everywhere, synapses misdirected, holes in my memory, gaps in the roof of time.

They say that we could hang turbines in the sky, kite-balloons lifting light petals of some sturdy substance that could turn wind, turn jet stream, turn spin into electricity, turn wheels, turn engines. No fossils. Only breath.

If only I could turn these rivers of lightening, this wild wind inside me, into a steady hum, no wild oscillation, no wobble, no saturated, unstable lift, nothing to turn breath into storm, trough into hurricane. Only breathing, in and out, the crackle calmed, the lovely, chaotic pulsing of fire making everything into poetry, branching through my body, blossoming in the slanted light of the world.

When I Write Wind

When I write wind I am writing the breath of the world, every single thing that breathes. I am writing *viento alisio*, breath of the sea, dance of heat and evaporation, chill and condensation, how temperature makes motion, and the sea wind, the wind that prevails, crosses vast expanses of salt swells, water heaving upwards to the sky and sinking back, rolling across the chasms of the deep, and I am writing storm surge and whitecaps, currents of air that twist and turn and shove the currents of water.

When I write wind, I am writing the gust that comes before the aguacero, flipping the leaves of yagrumo from green to silvery white, a wave of verdant light pouring down the mountain, wind made visible through the movement it shakes out of solid things. The trembling of aspen, the shiver of oaks, the rasp of palm fronds, the rattle of flamboyán, creak of bamboo, ache of cypress, great dunes of sand delicately rippled, and sandstorms that can grind stone, dull glass, erode bone into powder.

I write the racing of clouds, water lighter than raindrops, streaming along the courses of air, the rivers of oxygen taking their paths around the globe, and I am writing *move*.

I am writing the way that from the first gasping intake of air, breath cannot be still until death takes us, the way life is made of inhale, exhale, a billion exchanges of gasses entering and leaving us, ideas entering and leaving us, voices, loves, decisions, always in motion, that when we say to ourselves *be still*, we can only ever mean be slower. We are leaves trembling on the wind we ourselves have made. I am saying *don't hold your breath*. It won't keep you safe, only starved for air. *Movement is inevitable.* Breathe, I say, and feel the way breath enters us and oxygenates the flesh, the way it leaves us, taking away debris. I am writing the gusty sigh, the sharp inhalation, the uneven whistling of the dying, the sharp aliveness of air pushing into newborn lungs, returning as exultant, startled cry into the newness of every-thing. When I write wind, I write release, breathe, move.

FOUR

Clouds

Obatalá

We say we are the fulfillment of our ancestors'
dreams.
What dreams will we be the ancestors of?

Braided Prayer

1

Some of us want to burn up the world sooner than share it, in the iron-clad belief that there is not enough of anything, hoarding up life that withers when it's pulled apart like that, cursed to never feel sated, continuing to insist there is no consequence, even as it knocks on the door.

Some of us are all awake, feel each blow, our hearts like drums beating hard day and night, putting bodies and words in the path of unthinking greed, singing keep it in the ground, singing water is life, water is really truly life, singing life, painted banners flying.

Some of us doze in the shallows of comfort, unable to open our eyes, listening to the same song over and over, unable to make ourselves look on a future of smoking ruin, and so we never see the gardens in the smoke, putting one foot in front of the other, planning as if there is no precipice, because we can't imagine there is anything to be done.

2

There is the land where the water rises right to the edges of our mouths and we either flee or become amphibious, and there is the land where the leaves parch into tinder, waiting for the spark and the offshore winds to blow through miles of dry underbrush and turn trees into torches, houses into charcoal skeletons, and there is the land of spinning storms that can shatter cement, blow blunt objects right through walls, and drive the sea all the way to the foothills.

In one place the people pray for rain. In another they pray for dry land. In a third they pray for still skies and a different storm path. What if we could braid our prayers. Not my drought, your flood. Not let this storm curve into your part of the map instead of mine.

What if all the prayers became breath, became the one word: *listen.*

What if the engineers listened to the river, and not what they wish the river to do in the name of commerce. What if they listened to the whisper of water creeping beneath their ever-taller walls? What if they opened the gates of spring flood and let the river spread the nourishment of its sediment over the wide miles of swamp and let land rise up again out of the swirling waters?

What if the foresters listened to the first people who listen to the land and learned to set small fires against big ones, to make meadows for deer and burn away the kindling under the trees? What if they made a friend of fire?

What if we learned the best way to bend with the wind, listened to the trees that survive, deep roots, flexible trunks, and studied them, built roofs like the wings of gulls, went into the ground like malanga roots and let the great swirls of weather wash over us?

What if we listened to the earth itself, not the desperations of owner-ship, that fever that is never still? What if we listened to the drowning men on the crowns of the levees, and the women wading rivers bris-tling with guns, and the children of Amazonas watching the tree of life fall into the chipper?

What if we dig in deeper into the soil beneath our feet, and even when we stand on pavement grey, let ourselves hear lake water lapping with low sounds by the shore, what if we hear it in the deep heart's core?[2]

3

I am writing these words from the bed of an ancient sea, white expanses of dry salt, the fumes of copper smelters in the air, imagining how much listening would be enough to tip the balance, imagining

---

[2] *from The Lake Isle of Inisfree by W.B.Yeats,.*

what it would be like, a world of people who can feel the veins of copper in the earth, who hang prisms over the graves of plankton and do not disturb their oily rest, who hear the wind rising and know which note means time to take shelter, who make commerce change its course, not the river.

One of our teachers[3] says to learn the grammar of animacy, to know the aliveness of everything and speak to this quality by name. To know the water is being a bay, being a creek, being a rainstorm, is always being, that it was never, even in the dead of winter, an inanimate thing.

I am writing these words from a nest of bruises, the lightening in my head having hurled me at the steps and broken bone.

What if all the prayers became the one word: *listen*.

4

I hear the crows in the budding willow. I hear the guaraguao unfurl its high hawk's cry over the grassy slopes, I hear the water drift very slowly down the estuary toward the sandbar, and a big fly explore the crevices of the doorway, while cattle low across the water. I hear my heart lubdub lubdub. I hear moss growing on willow stump. I hear the water in the wet ground under the dead leaves. I hear beetles. I hear fluids creaking through my bruises. I hear my belly trying to digest the world. I hear oil rigs and trees falling and gunfire and the shuffling of paper designed for theft, but I listen deeper, deeper. I hear the heart of the world, its molten core, and all the seas washing all the shores.

5

We are the people of the tipping point, where the river meets the sea, where the rubber meets the road, and we could all die of habit while everything spins away, where life and death have we set before us

---

[3]   *Robin Wall Kimmerer in Braiding Sweetgrass*

and every single inhaled breath says *choose life* and every exhale grips tight to the usual and says *not yet.*

So what if we braid our prayers into the one word listen, and as we listen, bending toward the earth, waves of listening move outward like rings on water. What if people everywhere stop in their tracks and feel deep into the sound waves of the living world, and we see how the harm in our mindbodies echoes the harm in the world, sludge caking the vena cava and the Yazoo, feverish imbalance, immunity vanishing, cities and cells growing out of control, the cravings that devour us. What if the beauty of the world takes our breath and restores it, and we see with the inner eye of the heart how precious, how brief, how easily broken, follow each drop of water that circles the globe, in and out of rivers, in and out of lungs, in and out of bloodstreams, toda la sangre formando un río, making us one.

6

This is where the river meets the sea. Hush now. Hush. Listen.

When I write cloud

When I write cloud, I write the impossible rising, write shape shifting, evaporation and distillation, impurities left behind in the moment of shift, how we begin to float above what brought us here.

When I write cloud, I write great masses of vapor gathered into puffs and strands and thunderheads, tiny droplets too light to fall, suspended in mid-air, lit up from above by the sun, shadowing the earth below.

When I write cloud I am writing aspiration, lift, and I am writing the moment when we let ourselves be taken by the stream. I am writing pure possibility, beyond control.

There is a reason we see faces in the sky, see elephants and towers, horns and hands, flags and wild horses changing shape in slow motion, forming and unforming.

When I write cloud, I write becoming.

Weave

There is the warp of river bed and undersea ridge, a warp of edges, banks, walls, levees, coasts, and the weft of water always weaving back and forth, a flying shuttle of liquid purpose. The long weave goes round and round, spindles of mist and manoomin, ducks and barges, crawfish and peach pits, swamp grass and seagulls, pelicans and palms, granite and coral, cranberry and papaya, sometimes wispy, sometimes stout as rope.

We call the water lake, river, gulf, sea, rain, spring, as it moves through its forms, yarns of green and grey, indigo and turquoise, tan and chestnut, ochre and terra cotta, clear as glass, opaque as dirt.

I am in the high crest of the cordillera, looking south to the Caribbean Sea in the distance, white lines of surf drawn on many shaded blue, and cloud banks moving north and west across the land, while the ocean stays sunlit. Today there is a blizzard in the far north where the river twists through cities, under iron bridges, ice making a deep rim where only small, light things can walk.

The same water, a clear arc of seawater barely tinted green, a thin wash, curves up the sand and retreats. The tropical blue of the deeper waters lifts into the sky, goes transparent, clumps into grey, moves away north on the wind, to fall as snow.

So how could we ever be apart, we the people of this flowing, creek to billow, stem to stern, the water aching along its path, crying out to us in answer to our cries? Nothing that is between us is impermeable or permanent. The truth of water seeps through the fibers of our costumes, wetting us to the skin. We are wet to the skin and dreaming, naked and grieving, crackling with rage, at the edges of these waters.

I feel the tug of the river as it wells up from the pooling of rain far away north, and flows toward me here, with its long, long road, and everything it carries. It's a late morning in spring, and the clouds are gathering around these heights that are called Indiera, mist filled with voices of ancestors as the clay is full of their teeth, their shards, their memories like leaves time has turned to lace, becoming soil. The clouds are gathering. The rain is about to fall.

Will We?

What if we all become apprentice islanders, and learn the things that island people know? What if we learn to be flotsam, to be buoyant? What if we learn to wear tall boots and stay dry? Maybe there will be more ocean and less rain, or maybe the rain will just shift away. Will we become rain nomads, following the clouds like herds? Will we make dryland terraces held in place with old fishing nets and head scarves until the roots take hold? Will we carry water a bucket a time to sun bleached industrial tanks and make aquaponic gardens full of fish and water lilies that bloom under desert stars?

What if the people of the tidal zones, the floodplains, the estuaries and archipelagoes expand the ancient arts of rafting? What if we build towns that move with the swells and follow sea turtles along their sea roads? What if we move aside from the paths of hurricanes, anchoring far to the south until they pass, and return to gather the wood of fallen trees and dig our root crops from the muddy ground?

White men who claim to know things say we must abandon the water worlds and move to the polar continental bedrock, but what if we don't? What if we turn back toward the ancient salty home of all life? What if we move toward the rising water instead of running away? What if we lose our cities of steel and stone and become people who live in baskets? What if to save ourselves we float on baskets woven from the leftovers of extraction, plastic drums, polyester shirts, styrofoam coolers, nylon ropes, piled high with compost, and heavy with green growing corn, tomatoes, melons, morning glories?

What if the people of the plains make dreamcatchers that harvest water from the wind? What if we make windmills, dew traps, sun ovens from junkyard cars, tin cans, broken appliances? What if the people of the great inland prairies learn to call rainstorms, make basins and chutes and cisterns, and cherish every drop, lean into the prevailing weather, collect their snowmelt, and in the scorching sun of summer, make islands of shade that rise out of a sea of grass?

What if the people of the high mountains grow quinoa to trade for fish, and what if we turn flooded basements into fish ponds and ask mushrooms and sunflowers to help us clean the soil? *They say that we could hang turbines in the sky, kite-balloons lifting light petals of some sturdy substance that could turn wind, turn jet stream, turn spin into electricity, turn wheels, turn engines. No fossils. Only breath.* What if we did?

What if we befriend pond scum and make our fuels our plastics our waterproof boots from things that grow, and leave the ancient dead to dream in their fossil beds? What if we make our meatloaves from algae and wild leeks?

When the coastal cities begin to drown, will we turn the lower stories of skyscrapers into oyster beds and clam farms, seed coral onto steel frames, make offices into greenhouses, roof gardens into farms, and paddle our boats between towers? Will we make airy bridges between blocks of apartment buildings and live like spiders? Will we grow bee balm and hyssop in every crevice of the cracked concrete and tenderly restore the hives?

What if we never drown in despair, but learn to float?

What if here and now in the death throes of the age of greed, in the fevered ramping up of extraction, with our people falling around us, we crouch around the fire we keep making in a bed of shredded headlines, from the friction of our hopes rubbed like prayer beads between our palms until they spark, and tell each other these stories. What if here and now as the brute force of taking throws us up against the border wall of impossible, we use these stories like chisels to pry apart the stones?

What if these stories are paleochannels running wild under levies and borders, under the fields of monocultured, genetically modified, pesticided industrial crops, maroon rivers full of secret nutrients spilling into the sea of us to feed our plankton, and exhale waves of oxygen so we can breathe. What if these stories steam up from our mouths as we talk, and make big puffy clouds of possibility circling the globe and falling as sweet rain into our upturned thirsty mouths?

Everybody says what if we don't win, what if we don't stop the war in time, what if we can't save each other's lives, what if we can't flip the situation, what if we can't climb the hill of time and roll down the other side, what if we don't make it? Drum of my heart, drum of the world, nothing is lost, water returning, swamp fed, red blooded, shehechianu: the rivers are ancient, but the moment is new, so what if we do? What if we do? What if we do?

Epilogue: Water Road

1

The Río Cubuy runs over huge white, water-carved stones between steep, rainforested slopes in the sacred mountains of Yocahú, renamed El Yunque, the Anvil, by men used to beating steel into swords.

I am sitting in a shallow pool of the river, water like silk flowing around me, dissolving barriers. Birds cartwheel in the air above. Clouds gather around the green peaks. A distant waterfall moves silently down a gleaming rock face.

In the deep stillness, their voices well up. *Come home.*

Suddenly it seems obvious. The questions that have been beating their wings in my face for months, years, *where to be, where to be,* where can I find clean air, peace, wild lands, brown people, community—all that turbulence settles onto the water like a great white bird and floats calmly, making ripples that move outward toward the river's edge and disappear. Everything around me and within me says *come home.*

2

There are people who write articles identifying the safest places to be as the world comes apart. They say to settle on the continental bedrock, away from the coasts, closer to the poles. Someone tells me that Duluth, Minnesota is a prime spot for climate refuge. Someone else says people in the hot and humid US South should turn their eyes toward Canada.

This morning the news is full of Canadian flood waters, thousands evacuated from four provinces, two of them solidly inland, states of emergency in Montreal and Ottawa. Meanwhile, the tidal people of Bengal, where snowmelt meets sea and islands appear and vanish every few years, are building farms on rafts made of plastic debris, lashed together and heaped with fertile soil. Meanwhile in my own Caribbean homeland, people are planting root crops, experimenting with earth ship houses made of tires and dirt with round roofs that deflect hurricane winds. Meanwhile the storm and pillage driven exodus from Boriken is contradicted by a small, steady flow we call *rematriation*. Across the vast diaspora of our people, some of us are being called home. We come with seeds, tools, the street skills and friendships we acquired in those northern cities. We come packed with intent.

3

This piece of land where my heart unfurls has had many names. I only know a few. My young parents, blacklisted city kids turned farmers by necessity, married less than three weeks when US troops entered Korea, call it Finca La Paz, a place of peace. A decade later, Lencho Perez, the person whose hands go deepest into the soil, who cuts and clears and plants for my parents, calls it Monte Bravo, fierce mountain, and he should know. But Lencho is long buried, and the land has eaten our house, leaving only a faded tile floor, a water tank. The forest has reclaimed the tilled acres, making an island of trees between clear cut slopes of banana and short, sun grown coffee bushes scored with eroded gullies that wash the soil downhill toward the sea. The cool air that rises from among their leaves precipitates rain, calling it down from the clouds that blow over and through the cordillera. Their roots bind water into soil. Neither fierce nor peaceful, these thirty-three acres, these trees, insects, ferns, birds, orchids, flowering herbs have become a sanctuary, an encampment of non-human water protectors calling us to join them. We rename the land Finca La Lluvia, Rain Farm. As moisture gathers in the air, the little coquis, tiny tan frogs hidden in the cups of bromeliads and the mossy crevices of fallen logs, begin their chorus of *toa*, endlessly repeated. They call for the mother and are answered with rain.

4

My dreams are full of yautía, malanga, the root foods of the islands, tasting of earth, full of nutrients, harbored in the rich soil, out of reach of storms. Pumpkin vines wind through and around their stems. Words fly from my mouth in flocks of endangered butterflies, of cucuyos, the little fireflies, winking off and on in the dusk.

If the *patria* is defined by ownership, by flags and borders and treaties between states, then the *matria* is this: the densely woven kinships of the land, this single organism of roots and fungi, mist and pollen, anthills and circling red-tailed hawks, their feathers turned amber in the sun, centipedes and berries and the billions of microbes unmaking the dead in the soil. All this matted interdependence is made of water, the one liquid being to which we belong, and all the molecular codes of creation it carries.

5

The water I carry is also made of language, the stories we humans make the way plants make chlorophyll, because story is our nature. But the path of sunshine into leaf has no dead ends or detours, while the water of our mouths can be tainted. Our rivers move across battlefields and wastelands, are rainstorms passing through sulphurous clouds of smoke. Like the privatized water of mountain springs, they have been bottled in plastic, diverted into rusted pipes that leach lead. I must let the wellsprings of story that rise up in me pass through beds of limestone, be filtered through the bones of a billion tiny ancestors, evaporate in the heat of this warming world, be drawn back to earth by the cries of tree frogs and kinfolk. I must call out in my own voice, toa, toa, for my words to rain down, sweet and clean.

6

Everywhere on earth the matria cries out to us to come home, to weave ourselves back into a fabric that is torn, fraying, falling into pieces, dying of separation. Everywhere on earth, molecules of water move though skin and sky, tracing the flows that bind us, teaching us to be woven. As I turn toward the high cordillera, toward the teal and turquoise sea that crashes in white foam against the reefs and rocks of my archipelago, my dreams follow the infinite paths of water.

It is no longer intellect that makes the maps. I can taste traces of Mesabi iron in tropical rain, feel the thinnest possible film of slick river mud between my toes as black ferrous sand scrubs my toenails pink, just east of Cueva del Indio. I feel the movement of all these currents in the veins and pools of the earth and in my own watery body. From the River's northernmost rootlets to the cloud valleys of Maricao, from the vast estuaries of South Louisiana to the moonless beaches of Borikén where sea turtles nest, the path of water tugs and pushes at my heart, a living rope, a story in endless motion, or so we pray. The more I give way to its pull and let it carry me, the more I know myself to be silt, just a fistful of mud, a scoop of nutrients the current brought to shore, here for a moment, soon to wash away. I am the residues of every choice I made, of all my ancestors, their triumphs and defeats, carried by currents that circle the world again and again. How do I end a story of water that doesn't end? Like this.

I set out from Minnesota asking only to listen to the River and let it tell me where to go, and the river took me in its grip and sent me south and farther south until it brought me here. I set out to listen, and the River brought me home.

Gratitudes

Many people nourished this book and the journey it came from.

A Studio in the Woods has been nurturing art in conversation with the River's ecosystem since it was created from wood, clay and dreams by Lucianne and Joe Carmichael. Now owned by Tulane University, it's a sweet, wild patch of forest, nestled against the levee, studded with hand-built houses, a biological research station and incubator of art. This book was conceived in residency at the Studio in April, 2005 and born in a second residency in the fall of 2018.

I am deeply grateful to the Carmichaels for creating it, and to the creative and dedicated staff that keep it going: Ama Rogan, Cammie Hill-Prewitt, Grace Rennie and Dave Baker. Thanks also to Kathy Randels who was my official Artist Ally and helped me keep this project from expanding beyond control. My thanks to Ashé Cultural Arts Center and the New Orleans Center for the Gulf South for opportunities to share this work while it was in process.

In September, 2018 I journeyed down the course of the River from Minneapolis to New Orleans. This journey would not have been possible or a fraction as meaningful and fun without my travel companion and co-dreamer Naiomi Robles of Holyoke, Massachusetts. Together we waited for eagles, sang water songs, laid tobacco, followed butterflies and sunflowers, felt the burdens of history, laughed and cried, listened to and spoke with ancestors, and found the place where our own family trees tangled.

Many people gave us guidance, stories, and sustenance along the way, and continued to support my explorations of the estuaries of the lower Mississippi. Thanks to Sharon Day and Sandy Spieler in Minnesota, the rangers at Effigy Mounds National Monument in Iowa, river guide John Ruskey of Quawpaw Canoe Company in Clarkesdale, Mississippi, and Jamya Payne, young filmmaker of Ruleville, Mississippi, and to Mark Tilsen, Cherri Foytlin and all the other water protectors of Camp L'eau est la Vie in Lafayette, Louisiana. Thanks also to Jonathan Mayer, the Feral Possum, who shared the Studio and his own artistic process with me during my residency, for late night conversations about language, the cross-roads cultures of Louisiana and the Caribbean, art-making, recipes, monster stories, and a lot of laughter. Thanks to Jillian Meyers, who thinks about fish, and drove me "down bayou" to explore the water-laced edges of the land, along tiny roads lined with immense stands of grass and ghostly oaks, past channels choked with water hyacinth, floating bridges, shrimp boats and lift boats. Thanks to Jayeesha Dutta and Monique Verdin of Another Gulf is Possible, dreaming floating fantasies and sharing good food, and to Monique for her film *My Louisiana Love*. To Dr. Alex Kolker who talked with me about paleochannels, river management and coastal ecosystems, and Dave Baker, Studio biologist who showed me apple snail eggs, took me walking in wetlands and told me about what it really means to preserve forests.

Thanks to Rebecca Snedeker and Denise Frazier of the Center for the Gulf South for supporting my return to New Orleans in the spring of 2019, and for dreaming big.

Thanks to the New Orleans chapter of Jewish Voice for Peace with whom I had the privilege of working during both my visits to New Orleans, and who organized a public reading for me, fed me delicious food, gave me rides, found me housing, and made the best use of what I had to offer.

Thanks to Matt Shwarzman, Kathy Randels, Jayeesha Dutta, Patrick Staiger, Sister Alison McCrary and the Studio staff for connecting me into the relational web of their New Orleans worlds. Special thanks to Sister Allison McCrary and Patrick Staiger for making space in their homes for me.

Thanks to the members of my wider community who donated money for me to travel, with the support of my dear friend Susan Raffo, to my homeland, Puerto Rico, land of devastation and possibility.

My deepest appreciation to Susan herself, who always understands my dreams, and who came with me on the last leg of this particular journey to Puerto Rico, where among many other things, I found the ending of this book, and for sharing deeply thoughtful silences and conversations, bowls of papaya and massive amounts of fried fish and mofongo. Thanks also to Don Luis Ríos, one hundred years old and still spinning stories about the mountains we share, and to Angel and Milly of Casa Vista Mar in Manatí, Puerto Rico, who, when it turned out their Airbnb apartment was too scented for me, set us up in magnificent comfort under the stars, and to Rosa Pla for housing in Indiera. My gratitude to Guido Frosini, steward of True Grass Farms, for offering the land he tends as a place of refuge where, surrounded by grassland and wetland, egrets, vultures, hawks, crows, doves, owls and red winged blackbirds, I have been able to hear myself think and therefore, to write.

As always, I am grateful to my web of co-thinkers which, at this moment most actively includes Annie Burdett, April Rosenblum, Aya de León, Gwyn Kirk, Mauricio Abascal, Ricardo Levins Morales, and Ruth Mahaney; to Danny Bryck, Rebecca Pierce, and Sydney Levy, my Unruly comrades; to my co-counselors, Catalina Bartlett, Claudia Martinez, Jennileen Joseph, Julianne Gale, Joelle Hochman, Julie Saxe-Taller, Michael Saxe-Taller, Rosa Blumenfeld, the sedition support group, and the gang at Personal Counselors in Seattle.

Finally, I am deeply grateful that I was born in the right year, the right place, the right family, to be an artist in these times, rooted in the physical and cultural soil of places I love, threading my words through the tangled web of stories by which we feed each other and nourish imagination and hope.

How to Support My Work

Thank you for reading *Silt*. I am grateful for the residencies and other institutional support that made it possible for me to take the time to research and write it. But as a disabled, chronically ill radical elder, without the safety net of an academic job, I also need the community of those who read and use my work to financially support its creation.

If you pass my book on to someone else, buy a copy to send to a prisoner or a domestic abuse shelter or clinic waiting room or your rabbi. I don't want to stop the easy flow of my writing through networks of people who will make good use of it, but I do want to be paid for writing it.

For forty years I've been writing poems and essays that are widely used in our social justice movements, and almost all of that work has been unpaid. I have built up a good amount of social capital—my ideas and words are valued and respected, and people write to me to tell me how much the books and articles and poetry have meant to them—but this hasn't translated into the income I need in order to pay my bills and have a healthy, productive old age.

I am experimenting with a new economic model, in which the broad community of people who read, teach, and quote my work pays me for the *practice* of continuing to bring my particular perspective to the challenges of our times, pays me to write and speak and broadcast what I have to offer, rather than waiting to buy my products. This frees me to say the things I think need saying, without having to package them into a book I can sell. It frees me to use my limited energy to tell the stories that need telling, instead of chasing grants and spending my time marketing. It frees me to accept only the speaking engagements that excite me, protecting my time and my health. It lets me give away books to those who can't afford them, without losing my livelihood.

Patreon is a funding platform created by and for artists. You join by pledging a small monthly amount. This brings you into an inner circle of people who actively back me to do what I do. It also means that you receive regular posts from me, in which I let you know what I'm working on, comment on what's happening in the world, and share excerpts of unpublished writing, and you can ask me questions, comment on my posts, and engage other supporters in conversation.

So please share my work. Buy copies for your friends and relations. Assign the whole book to your classes instead of a single excerpt for which I will never receive royalties. But I also urge you to join my Patreon community and help me to write the next book, and the one after that. Go to *https://www.patreon.com/auroralevinsmorales*. I hope to see you there.

Aurora Levins Morales
www.auroralevinsmorales.com

www.ingramcontent.com/pod-product-compliance
Lightning Source LLC
Chambersburg PA
CBHW031146090426
42738CB00008B/1236